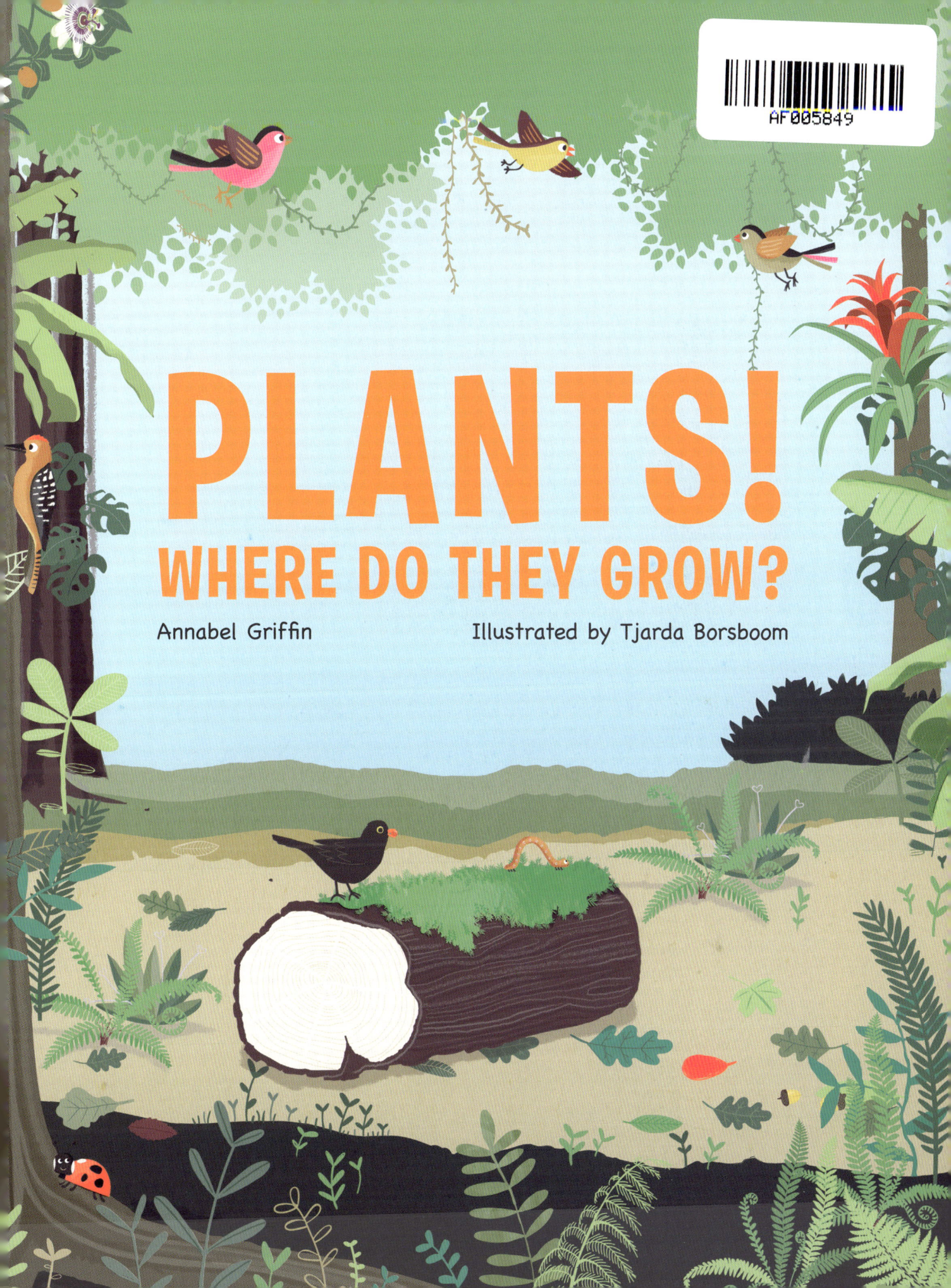

Contents

What Is a Plant?	4
Survival in the Desert	6
Beauty in the Rainforest	8
Peaceful Pond Plants	10
Fragile Forests	12
Adapting to the Arctic Tundra	14
Magnificent Meadows	16
Wonderful Plant Habitats	18
Did You Know?	20
Match Up the Pairs	22
Glossary	24

Words in **BOLD** can be found in the glossary.

Copyright © 2024 Hungry Tomato Ltd

First published in 2024 by Hungry Tomato Ltd
F15, Old Bakery Studios, Blewetts Wharf, Malpas Road, Truro, Cornwall,
TR1 1QH, UK.

No part of this publication may be reproduced, stored in a retrieval system, or transmitted in any form or by any means, electronic, mechanical, photocopying, recording, or otherwise, without prior written permission of the copyright owner.

A CIP catalogue record for this book is available from the British Library.

ISBN 9781916598881

Printed in China

Discover more at
www.hungrytomato.com

Picture Credits:
Abbreviations: m-middle, t-top, l-left, r-right, bg-background.

Shutterstock: Dennis Van De Water 18bm, 22ml; Gerrit Lammers 23bl; hor Winje 23mr; Ivan Kovbasniuk 21bl; Joshua Haviv 18ml, 22bl; Ksenya_89 19mr, 22br; Kwintek7 22tl; Lois GoBe 23br; MasterQ 19t, 22mr; Mats Brynolf 23ml; Oleg Kovtua Hydrobio 23tl; Petr Baumann 23tr; S.tomas 21tl; Thorsten Schier 19bl, 22tr.

Every effort has been made to trace the copyright holders, and we apologise in advance for any unintentional omissions. We would be pleased to insert the appropriate acknowledgements in any subsequent edition of this publication.

What Is a Plant?

Plants are living things that can be found almost everywhere on Earth! There are over 300,000 different types of plants on our planet. How many can you name?

Plants come in all sorts of shapes and sizes, but most of them have the same three parts: stem, roots, and leaves.

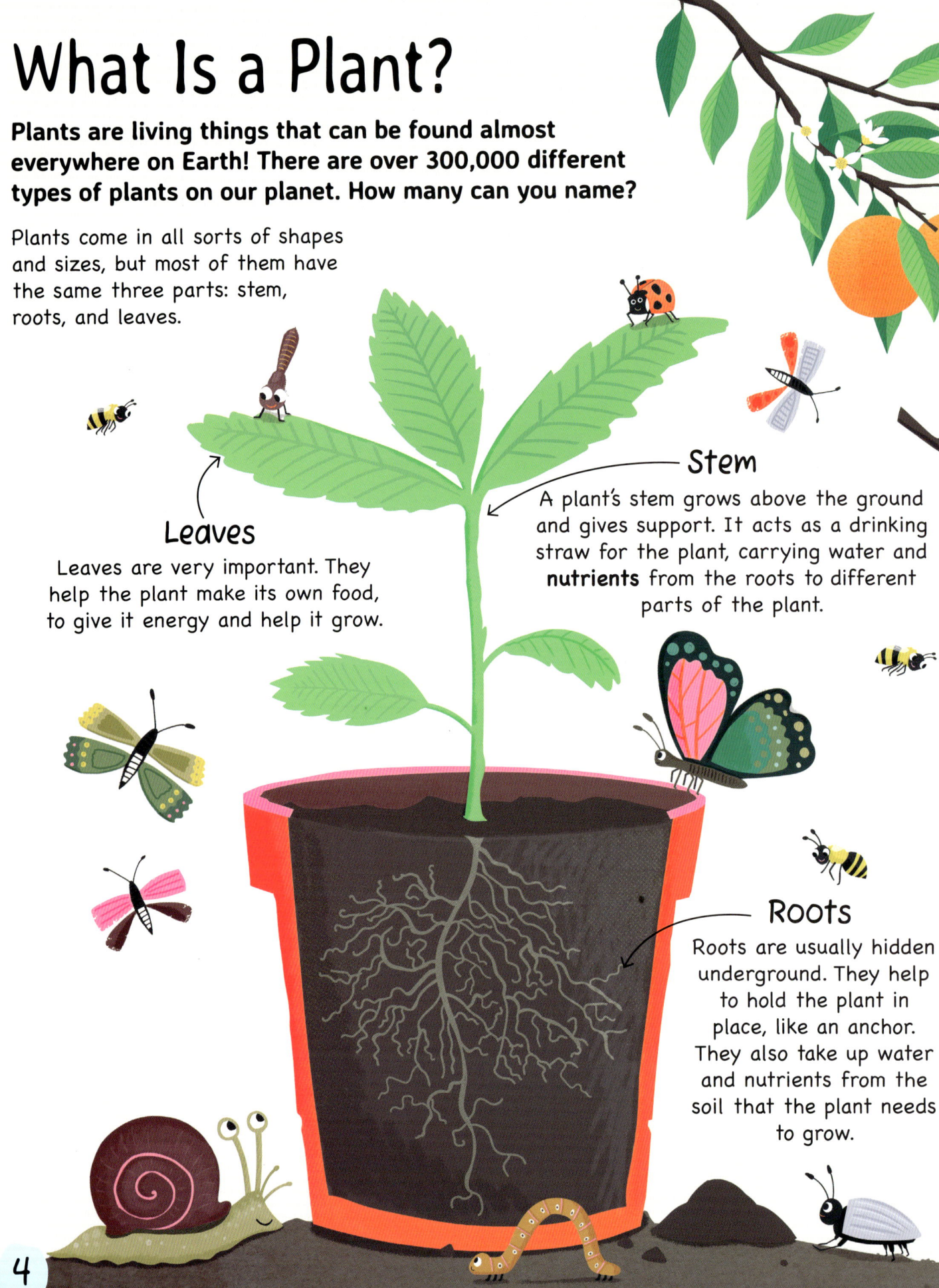

Leaves
Leaves are very important. They help the plant make its own food, to give it energy and help it grow.

Stem
A plant's stem grows above the ground and gives support. It acts as a drinking straw for the plant, carrying water and **nutrients** from the roots to different parts of the plant.

Roots
Roots are usually hidden underground. They help to hold the plant in place, like an anchor. They also take up water and nutrients from the soil that the plant needs to grow.

Survival in the Desert

Deserts are hot, dry places, where there is very little rain. Some desert plants, such as cacti and agave, can store precious water in their stems or leaves; other plants must find different ways to get water.

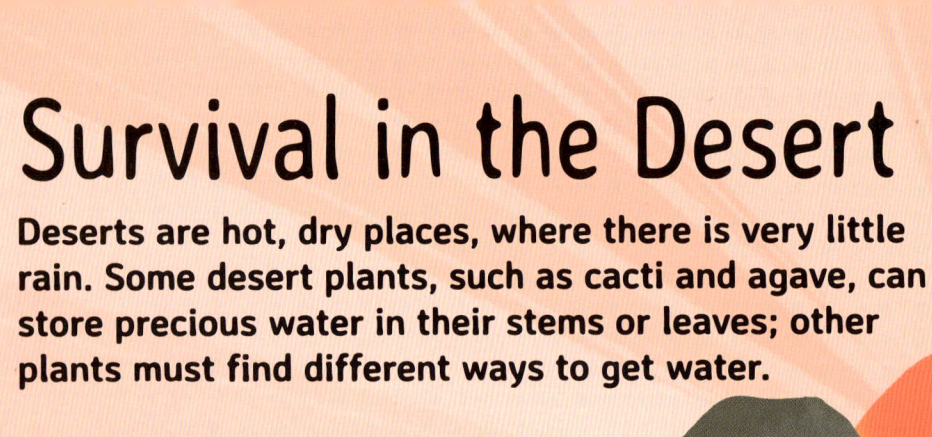

Tumbleweed
When a tumbleweed is fully grown, it separates from its roots and rolls away in the wind, spreading its seeds as it goes.

Please stop following me!

Water underground
Desert plants often grow very long roots that spread deep underground to find water.

Beauty in the Rainforest

Tropical rainforests are home to thousands of different plants. Lots of different animals, including monkeys and birds, make their homes high up in the trees.

Lost in the forest

Scientists believe there may still be thousands more rainforest plants that haven't been discovered yet!

Tropical fruit

Tasty fruits like bananas and passion fruit grow in the rainforest.

Cacao pod

Cacao trees grow large **seedpods** full of cacao beans, which are used to make yummy chocolate!

Orchids

There are over 25,000 different types of orchid, and most of them grow in tropical rainforests.

Passion flower

Passion flowers are some of the strangest and most beautiful flowers to grow in the rainforest. They also produce tasty passion fruit!

Bromeliads

Bromeliads are plants that often grow high up on the branches of trees. Their leaves collect and hold rainwater, like a bucket. Poison dart frogs sometimes lay their eggs in the little pools of water.

Water lilies have wide, flat leaves that float on the surface of the water. Their roots reach down to the bottom of the pond.

These plants have swollen, spongy stems which are full of air to help them float. Their roots float freely in the water.

Underwater plants like hornwort can provide safe hiding places for fish and newts.

Cattails, or bulrushes, are a type of reed that grows at the water's edge. They have sausage-like flowers.

Fragile Forests

Forests are large areas of land that are covered in trees. They are home to lots of different types of wildlife.

Tree houses
Some owls, foxes, squirrels, and other animals make their homes inside holes in trees.

Fun fungi
Mushrooms and toadstools can often be found growing in the forest. These are known as **fungi**.

Green planet
Different kinds of forests exist all over the world and cover about a third of the land on Earth!

Spring and summer

Trees can look very different depending on the season. In spring and summer, many trees grow lots of green leaves all over their branches.

Autumn

Many trees go through a big change in autumn. Their leaves change from green to red, yellow, orange, or brown. They also start to lose their leaves.

Winter

By winter, most tree branches are bare. Trees that lose their leaves in winter are called deciduous trees.

Some trees stay green all year round. They are called evergreen trees.

Adapting to the Arctic Tundra

The Arctic tundra is one of the hardest places for plants to grow in. It is very dark, cold, windy, and dry there.

Snow and ice covers the ground for up to 10 months of the year, and most of the soil remains frozen all year round!

There are no large trees in the tundra. Tundra plants are small and grow close to the ground. Most grow in the short summer, when the snow has melted.

In the winter, the temperature can fall to a bone-chilling -70°C (-94°F).

Many animals **hibernate**, or move to warmer areas, during the long, cold winters.

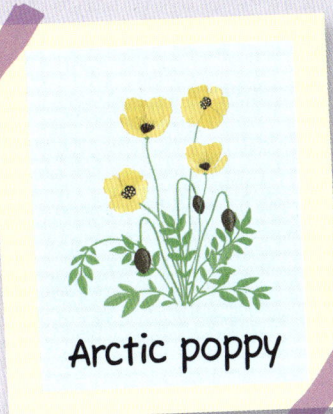

Arctic poppy

The flowers of the Arctic poppy move their faces towards the sun to soak up warmth and light.

Moss campion

Moss campion grows in cushion-like mounds. This helps it hold on to warmth and moisture.

Arctic cotton grass

Cotton grass has thin stems with fluffy cotton-ball tops. The fluffy balls contain tiny seeds, which are spread all over the tundra by the wind.

Bearberries

Bears love to snack on these berries.

Arctic bell heather

These little flowers have tiny leaves that overlap, like scales on a snake.

Arctic willow

The leaves of the Arctic willow are covered in long, fluffy hairs, to keep it warm.

Magnificent Meadows

A meadow is a large area of land that is mostly covered in amazing grasses and wonderful wildflowers.

Hay there!
Meadows can be used to grow hay, which is used to feed farm animals in the winter.

Meadow munching
Farm animals, like cows and sheep, often graze in meadows.

Buzz buzzing
Pollinating insects, like bees and butterflies, love wildflower meadows!

Grasses

Wildflowers

About a quarter of plant life on Earth are grasses. They play an important role in keeping our planet healthy.

Wildflowers grow naturally in the wild. They are beautiful and attract wildlife, such as insects and birds.

Make Your Own Wildflower Seed balls

Turn a patch in your garden or outside space into a mini meadow.

You Will Need:
- Soil or peat-free compost
- Flour
- Packet of wildflower seeds
- Water

1. Mix together 85g of compost, 28g of flour, and a packet of seeds in a mixing bowl.

2. Add small amounts of water and mix with your hands until everything sticks together.

3. Roll the mixture into small balls and leave to dry in the sun.

4. Once dry, throw your seed balls into empty flowerbed spaces.

Wonderful Plant Habitats

Our world is made up of different habitats, each one home to specific plants, animals and weather. We've learned that plants can grow in extreme habitats, but what makes each one so different?

Deserts

Deserts can be very hot or freezing cold. In fact, the world's biggest desert is the frozen desert in Antarctica. Whether in the hot or cold, all desert plants need to **adapt** to the dry environment.

Deserts get less than 25cm of rain every year.

Small plants that grow low to the ground.

Rainforests

Rainforests are home to over half of the world's plant **species**! Rainforests don't have seasons; they have the same weather all year round.

Rainforests are hot and humid.

Trees grow very tall to fight for sunlight.

The rainforest has lots of plants, animals and bugs.

Aquatic habitats

Aquatic habitats include oceans, lakes, ponds, and rivers which makes them the biggest of all! It's not easy for plants to live in water, so aquatic plants are very special. Most live in either freshwater or saltwater, few can survive in both.

Plants here provide food and shelter for animals.

Plants can live in, on top of and around water.

Arctic tundra

Arctic tundra is found in the far north of the world. The soil has few nutrients which means that not many plants can grow there. Those that do are at risk from **climate change**; as the planet warms up, the Arctic tundra shrinks, and plant numbers decrease.

Cold, dry weather all year round.

Plants grow in groups to stay protected from the icy wind.

No trees in sight!

Temperate Forests

Unlike rainforests, temperate forests have 4 seasons a year. In summer, the trees have green leaves which slowly fall leaving branches bare in winter. Dead leaves break down, adding nutrients into the soil, making it a good place for new plants to grow.

Leaves change through the seasons.

Forests are full of different trees.

Did You Know?

Plants are pretty amazing! Every living creature needs plants to survive; the world wouldn't be the way it is today if we didn't have them. Did you know these amazing facts about plants?

Not all cacti grow in the desert – some grow in the jungle, mountains, and even on your windowsill at home!

It is thought that there are more trees on Earth than stars in the Milky Way galaxy!

Rainforests are Earth's oldest living **ecosystems**. Some are more than **70 million** years old!

Bolivian waterlily

The biggest aquatic flower is the Bolivian waterlily whose leaves can grow up to **3.2 metres.** That's wider than most buses!

Scientists discover around **2,000** new plant species EVERY YEAR!

NEW SPECIES · WHAT'S THAT PLANT? · PLANTS AROUND THE WORLD

Forests cover about **30%** of the land on Earth. This includes temperate forests, tropical rainforests and frozen forests!

Match Up the Pairs

Can you match up the photographs of habitats (below) with the plant that lives there (right)? Flip back through the book if you need a hint!

1. Meadows
2. Forest
3. Rainforest
4. Ponds/aquatics
5. Desert
6. Arctic tundra

A.
Cactus

B.
Water lily

C.
Arctic cotton grass

D.
Fungi

E.
Passion flower

F.
Wildflowers

Have you matched them all?
Answers can be found on page 24.

Glossary

Adapt – (verb) when a living thing is able to survive in its surroundings by developing special features or skills over a long period of time.

Aquatic – something that grows, lives or spends a lot of time in water.

Climate change – long-term changes in temperature and weather.

Ecosystem – an area where living things work together to form a physical environment.

Fungi – (the plural of fungus) A group of living things, including mushrooms, moulds and yeasts, that are neither plants nor animals. They reproduce using spores instead of seeds, and don't produce their own food, like plants.

Habitat - the natural homes of plants and animals. Habitats have different characteristics that affect which plants can grow there.

Hibernate – (verb) to go into a deep sleep-like state for a long period during winter, when there is little food.

Nectar - a sugary fluid made by flowers to encourage pollination (see below).

Nutrients – substances or ingredients that plants and animals need to live and grow.

Pollination – when pollen is moved from one plant to another – often by an insect – so the plants can make new seeds.

Seedpod – a pouch or case produced by some plants to hold their seeds.

Species – a group of living things that are the same as each other. For example, African forest elephants and African bush elephants are two different species.

Trunk – the large woody stem of a tree, where the branches grow from.

Answers to Match Up the Pairs

1. Meadows - F. Wildflowers
2. Forest - D. Fungi
3. Rainforest - E. Passion flower
4. Ponds/aquatics - B. Water lily
5. Desert - A. Cactus
6. Arctic tundra - C. Arctic cotton grass